T·H·E
BIG
A·N·D
LITTLE
ANIMAL
B·O·O·K

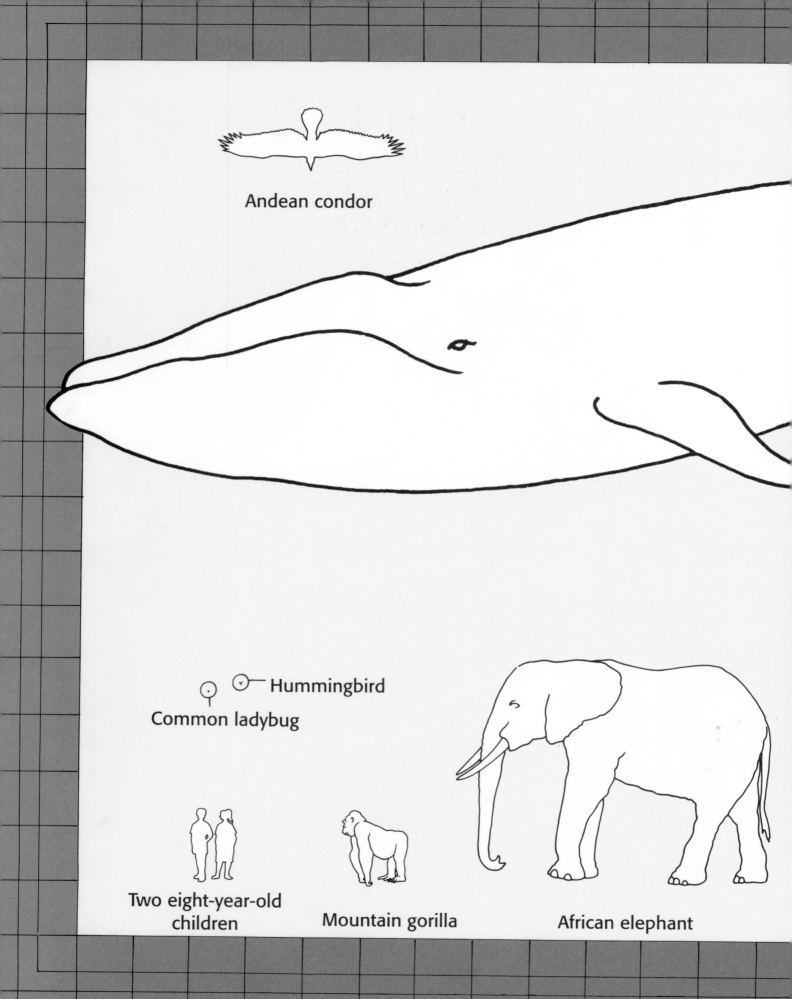

Andean condor

Hummingbird

Common ladybug

Two eight-year-old
children

Mountain gorilla

African elephant

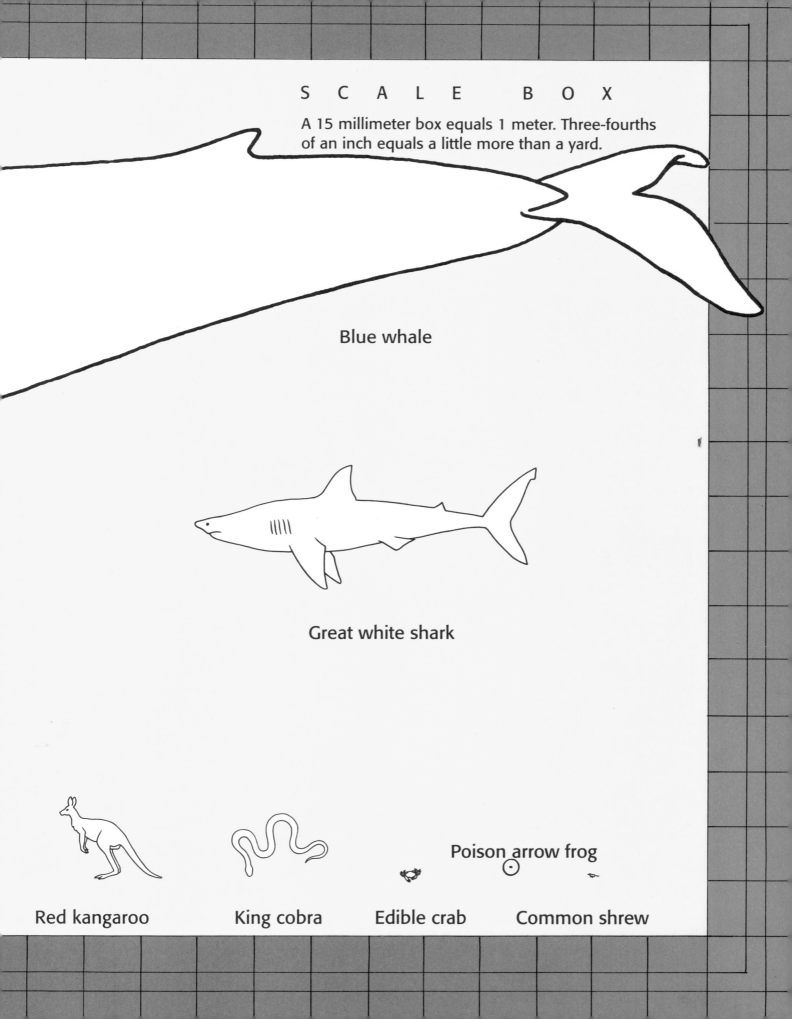

A 15 millimeter box equals 1 meter. Three-fourths of an inch equals a little more than a yard.

Blue whale

Great white shark

Poison arrow frog

Red kangaroo

King cobra

Edible crab

Common shrew

Animal Opposites

T·H·E BIG A·N·D LITTLE ANIMAL B·O·O·K

DAVID TAYLOR
ILLUSTRATED BY PETER MASSEY

RSVP

RAINTREE
STECK-VAUGHN
PUBLISHERS
The Steck-Vaughn Company

Austin, Texas

Contents

Words found in **bold** are explained in the glossary on page 31.

To Louise Coles, whom I measured.

Published by Raintree Steck-Vaughn Publishers, an imprint of Steck-Vaughn Company

Illustration copyright © Peter Massey 1994
Text copyright © David Taylor 1994

Editor: Jill A. Laidlaw
Science Editor: Tracey Cohen
Designer: Frances McKay, Julie Klaus
Electronic Production: Scott Melcer

Library of Congress Cataloging-in-Publication Data

Taylor, David, 1934–
    The big and little animal book / David Taylor ; illustrated by Peter Massey.
        p.   cm. — (Animal opposites)
        Includes index.
        ISBN 0-8172-3950-2
        1. Body size—Juvenile literature. 2. Animals—Juvenile literature. [1. Size. 2. Animals.]
    I. Massey, Peter, ill. II. Title. III. Series: Taylor, David, 1934– Animal opposites.
    QL799.3.T38   1996
    591—dc20   95-8332
                                                  CIP   AC

Printed in Hong Kong
Bound in the United States
1 2 3 4 5 6 7 8 9 0 LB 99 98 97 96 95

The publisher wishes to point out that children are included in the illustrations to provide a sense of scale. Obviously, children should not come into contact with dangerous or poisonous animals.

Big and Little Animals

The world is full of billions of animals that come in millions of wonderful kinds. There are those that live deep in the water and others that live on land, under the ground, or in the air. Some are too small to be seen, some are giants. Some are shy, dull-colored, and gentle. Some are bold, brightly colored, and fierce.

Different or the Same?

All animals have bodies, and all bodies have certain things in common. Animal bodies are made up of hundreds of millions of tiny living building blocks called cells. There are many kinds of cells, usually arranged in groups called **tissues** and **organs**. Simple animals, like jellyfish, have tissues such as muscle and **nerve**. In more complicated animals, tissues are grouped into organs, including a heart, brain, eyes, stomach, **liver**, and **kidneys**. Each kind of tissue or organ has a specific job, like trapping and digesting food, or getting rid of wastes.

Even though two animals may be different, they may have the same kind of tissues or organs doing the same kind of job. The giant blue whale has powerful muscles that are made of cells like those in the muscles of a housefly's legs. A mouse's heart is built very much like the heart of a boy or girl. An octopus eye has the same basic design as the eye of a tiger.

Where in the World?

You can see the places where the animals talked about here live. Turn to pages 30 and 31 in the back of this book.

Size is important and being big or little brings both advantages and disadvantages. Big animals need more food and more space to live in. Big animals can store their body heat better, live longer, and defend themselves more readily than little ones. Little animals find shelter more easily. Little animals risk early death by being eaten by other animals. When we use the words "little" or "big," what exactly do they mean? A flea would say that a mouse is big, and the mouse would be very little to a rhinoceros. It is the same when we compare things. A distance that is long to you or me may be short to a trained runner. We need more correct ways of explaining what we mean. That is why, thousands of years ago, people began measuring.

How Do We Measure?

To measure you need rules that are based on something unchanging that can be referred to. Different people, at different times all over the world, have invented their own ways of measuring. These measurements have been based on different things, such as the weight of a particular seed or the length of a person's arm.

When we measure length, we use inches, feet, yards, and miles. If you look at the ruler at the bottom of these pages you will see inches marked on it. Each inch is divided into smaller units—halves and fourths of an inch. There are 12 inches in a foot and three feet in one yard. You will also see metric measurement on this ruler and these pages.

A 3/8 inch box is equal to a little over a yard.
A 10 millimeter box is equal to 1 meter.

These scale boxes give an idea of how big and little the animals in the book are compared to each other. Two eight-year-old children are also included.

When we want to know how heavy something is, we use ounces, pounds, and tons to measure its weight. An ounce is very light, but a pound is heavier since it is made up of 16 ounces. A ton is as heavy as 2,000 pounds. This book weighs 13 ounces, which is the same as 185 bee hummingbirds.

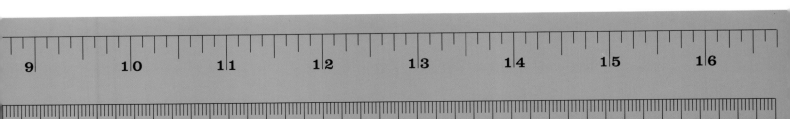

The Blue Whale

One of the biggest animals that has ever lived on this planet is still around. Or at least some are still around. It is our fellow **mammal** the mighty blue whale. The whale can grow to 108 feet (33m) in length. This is as long as the whaling ship of the 19th century. Those were the ships that went out to hunt whales! The same whale can weigh over 160 tons (145mt), or as much as two heavy jet-engine airplanes.

Yet this gigantic animal is a gentle character, quite unlike its smaller relative the killer whale.

Toothless Whales

Some whales, like killer whales and pilot whales, have teeth, but blue whales and other so-called baleen whales do not. Instead, they have rows of baleen plates in their mouths. Baleen, which was once called "whalebone," is not bone at all. It is made of horn—the same material as hair, hooves, and fingernails. Toothless whales use their baleen to sift krill, small thumb-length shrimp, out of the water. Krill is plentiful in areas like the Southern Indian Ocean where parts of the water are cold and rich in oxygen.

Blue whale

A Protected Animal

In the 19th century, there were more than 200,000 blue whales. They were hunted almost until they became extinct. About 12,000 survive, mostly in the Southern Hemisphere. They are protected by international laws.

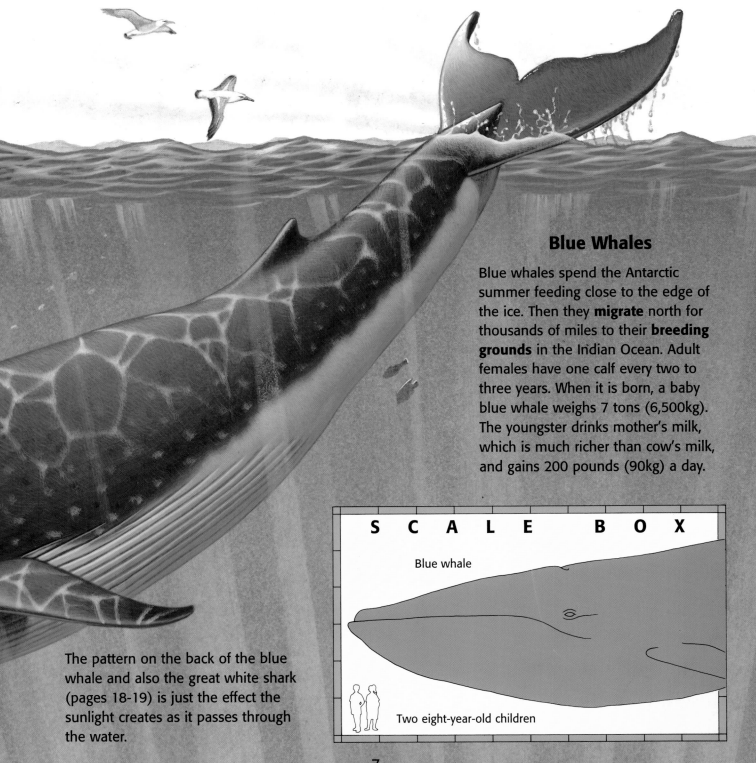

Blue Whales

Blue whales spend the Antarctic summer feeding close to the edge of the ice. Then they **migrate** north for thousands of miles to their **breeding grounds** in the Indian Ocean. Adult females have one calf every two to three years. When it is born, a baby blue whale weighs 7 tons (6,500kg). The youngster drinks mother's milk, which is much richer than cow's milk, and gains 200 pounds (90kg) a day.

S C A L E   B O X

Blue whale

Two eight-year-old children

The pattern on the back of the blue whale and also the great white shark (pages 18-19) is just the effect the sunlight creates as it passes through the water.

The Gorilla

If we are talking about big, then *King Kong* the famous Hollywood monster gorilla was really big. But, of course, he was imaginary. Gorillas are big—but not big enough to snatch an airplane out of the sky. That's what happened in the movie.

In fact, gorillas are best compared to human beings. A male gorilla in the wild weighs between 300 and 600 pounds (135 to 275kg). The record weight for a wild gorilla is 483 pounds (219kg). These weights are heavier than those of the average grown man, but Konishiki, the champion **Sumo** wrestler in Japan, weighs an incredible 555 pounds (252kg). Gorillas are far stronger than humans.

SCALE BOX

Mountain gorilla

Gentle Giants

Despite the King Kong reputation and the males' habit of beating their chests with their fists, gorillas are very peaceable animals. They live in family groups that stay together for years. These groups travel through the forest feeding as they go on plants, herbs, and fruit when it is available.

Animals at Risk

Gorillas are highly endangered creatures. The threat comes from their very dangerous relative—humans. Recently, *Mrithi*, the 23-year-old, 397-pound (180kg) star of the film *Gorillas in the Mist*, was shot dead by soldiers fighting a civil war in Rwanda, in Central Africa.

Mountain gorilla

The Hummingbird

Put three salted peanuts into the palm of your hand and feel their weight. They weigh about 1/14th of an ounce (2g), the same as a bee hummingbird. This bird is the tiniest bird in the world. The largest living bird, the African ostrich, weighs about 75,000 times more. Even a small house sparrow weighs at least twice as much as a bee hummingbird.

There are about 321 different kinds of hummingbirds. They have body feathers that are brightly colored and shiny, like polished metal.

Violet-cheeked hummingbird

Too Fast to See

Hummingbirds' wings are built something like your hand. But, instead of lots of fingers, they have one very long bone that could be a finger bone and two short bones. Hummingbirds' wings are different from other birds' wings because they swivel at the shoulder. This makes it possible for them to move in a figure-eight pattern. This extra movement means that a hummingbird's wing can beat up to 80 times every second. To us this movement is so fast that it looks like a blur in the air.

Food for the Birds

Hummingbirds live in the Americas and the Caribbean. They feed on nectar and small insects. Nectar is the sweet liquid found inside some flowers. Each kind or species of hummingbird has a slender bill. The bill is the perfect length and shape for getting into the type of flower that a hummingbird likes to feed off.

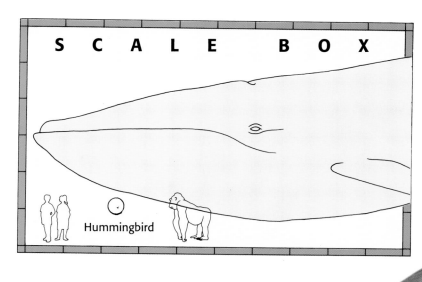

Flying Backward

Hummingbirds are like living helicopters. They are able to hover in midair while they feed on a flower. These birds can even fly backward, unlike any other bird.

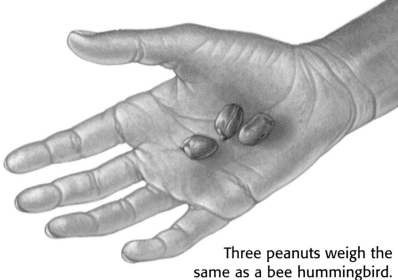

Three peanuts weigh the same as a bee hummingbird.

Bee hummingbird

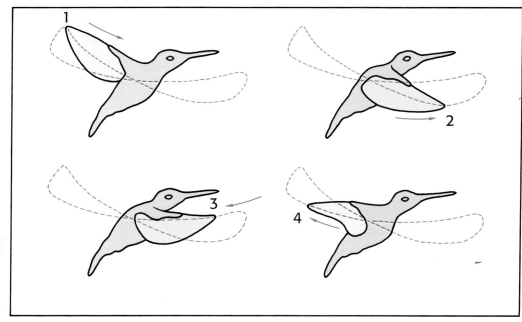

11

The Andean Condor

The amazing Andean condor is one of the biggest flying birds in the world. It is a bird of prey or hunting bird with a short, hooked beak, strong feet, and sharp claws. Condors are also vultures. There are seven kinds of vultures in North and South America.

Andean condor

While the California condor is now very rare, the Andean condor is still common. The Andean condor can still be seen soaring above the snowy peaks and windswept valleys of the Andes Mountains of South America. Male Andean condors weigh between 20 and 24 pounds (9 and 11kg).

Huge Wings

The wingspan of the Andean condor is just under 10 feet (3m). Wingspan is the distance between wingtips with wings extended. This is not quite as wide as the wingspan of the wandering albatross, which is just over 10 feet (3m). That is the width of three eight-year-old children standing in line with their arms outstretched and fingers touching.

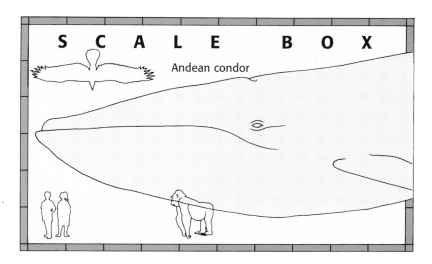

SCALE BOX

Andean condor

The Andean condor has the same wingspan as two eight-year-old children standing together with their arms outstretched.

13

The Poison Arrow Frog

Tiny, most beautifully colored, and highly poisonous—these are words used to describe the poison arrow frog. The poison arrow frog is found in Central and South America and parts of the Caribbean.

All frogs and toads make poisons, but most are not that strong. The poison of poison arrow frogs of Central and South America is strong enough to kill large animals. South American Indians use the poison on the tips of hunting arrows. One of the most dangerous is the Kokoi of Colombia. One frog makes enough poison in its skin for 50 arrow tips.

Poison arrow frogs range in size from .8 to 2 inches (2 to 5cm) long. Their bright colors warn predators of their poisons.

Poison arrow frog

Boulenger's toad (a type of poison arrow toad)

14

Land and Water

Frogs, and their close relatives the toads, are amphibians. Amphibians are animals with backbones that live both in the water and on land. There are about 2,000 different kinds of amphibians known.

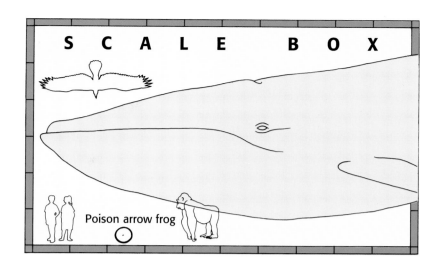

SCALE BOX

Poison arrow frog

Being Born

Most amphibians lay eggs without shells in water. The hatchlings, tadpoles, breathe underwater through gills. As they grow legs and are able to move onto land, many amphibians lose their gills. They develop lungs to breathe air.

The actual size of a golden poison arrow frog.

Breathing Through Skin

Frogs have thin, moist skin with many blood vessels and no scales. They use their skin to take in oxygen from the air. Frogs also get oxygen by using their lungs to breathe.

The African Elephant

The most gigantic animals that have ever lived on Earth were some of the dinosaurs. They were huge reptiles that could measure longer than fourteen men lying head to toe on the ground. They weighed as much as an empty jumbo jet.

Speaking of jumbos—now that the dinosaurs no longer exist, the biggest land animal around is the elephant. Elephants are intelligent animals who keep in touch with one another by using smell, sound, touch, and sight. Elephants are somewhat near-sighted and have thick, quite sensitive, ticklish skin!

Tusks as Teeth

Elephants only use four of their teeth at any one time. Each of these teeth is as heavy as a house brick. Elephant tusks are actually very big teeth. They can grow to 6 feet (2m) in length and weigh 20 pounds (53kg) each.

African elephant

Useful Trunks

Elephants can use their trunks both with great strength and very gently. They can tear down a small tree or lift a bird's egg without breaking it. An elephant's trunk can break off trees, dig holes, and pick up small objects. It can hold 15-20 quarts of water.

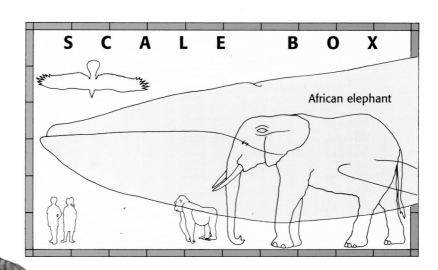

S C A L E   B O X

African elephant

Empty Heads!

Elephants have massive heads that contain many air spaces called **sinuses**. These sinuses cut down the weight of their heads.

There are two kinds of elephants—the African and the Asian. Adult African males can grow to 13 feet (4m) measured at their shoulders. This is taller than two men, one standing on the other's head!

The Great White Shark

Jaws! Who hasn't heard of the film that made the great white shark famous? It is one of the most dangerous creatures in the ocean. This scary beast is the third largest of all fish. The great white shark grows up to 20 feet (6m) in length.

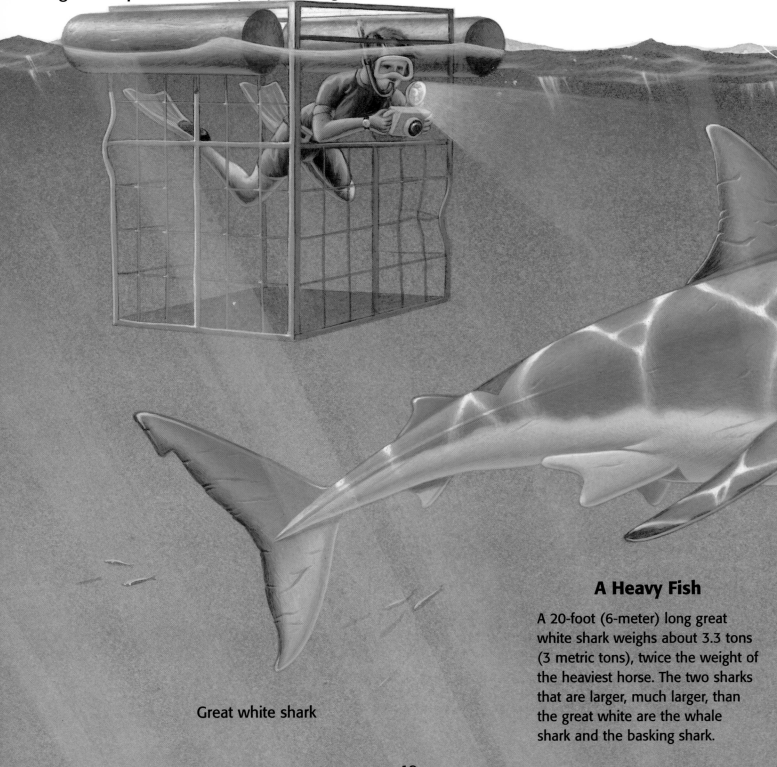

Great white shark

A Heavy Fish

A 20-foot (6-meter) long great white shark weighs about 3.3 tons (3 metric tons), twice the weight of the heaviest horse. The two sharks that are larger, much larger, than the great white are the whale shark and the basking shark.

This total length is longer than three tall men laid end to end! The body of the great white shark is gray-brown or gray-blue on top and off-white underneath, usually with a black spot on the armpit of each big side fin. This shark has a blunt, cone-shaped head. Its mouth is full of massive, jagged teeth shaped like triangles. It can be found in all oceans, but it prefers warm water.

SCALE BOX

Great white shark

The Shrew

Common shrew

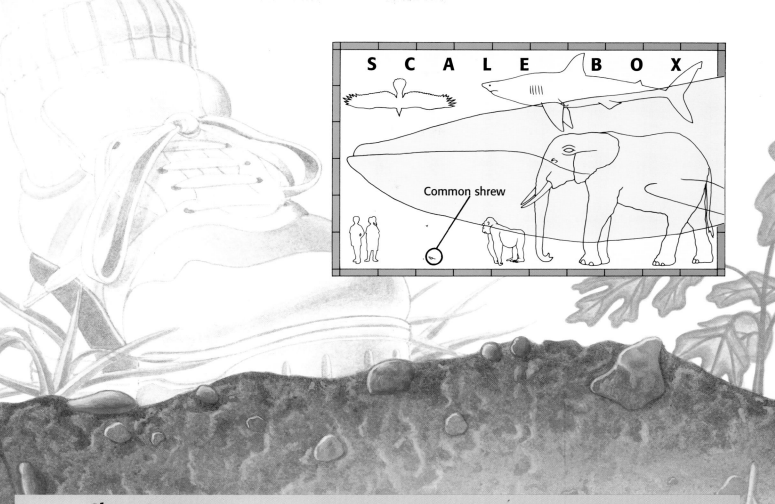

SCALE BOX

Common shrew

Shrew Language

Shrews have poor eyesight and a limited sense of smell. But their hearing and touch are very sharp. So they communicate by sound and feel. They talk to each other with high-pitched screeches and chirpings.

Dying Young

The forest or mouse shrew is one mammal whose life is very short. It dies of old age when it is just about one year old. The common European white-toothed shrew has a life span of three years.

Big Appetites

Shrews are busy individuals that are always looking for food. Most eat worms, grubs, and insects, but some also like fish, frogs, lizards, and mice.

A shrew is a shy little mammal that looks like a mouse but has a longer, pointed nose. The shrew has traditionally been known for its bad temper.

In the Woods and Water

There are at least 268 different kinds of shrews spread widely across the world. Most live in the undergrowth of forests. There are some water-loving kinds that hunt in streams and rivers. Shrews range in size. The Madagascar pygmy white-toothed shrew measures about 1.8 inches (46mm) long, not counting its tail. It weighs just 0.07 ounces (2g), as much as three raisins. The African giant shrew reaches 5.6 inches (140mm) in length and 2.3 ounces (65g). The pygmy white-toothed shrew is the tiniest land-living mammal. It is so small it can crawl through the tunnels left by earthworms.

The Kangaroo

In 1770, Captain Cook first landed in Australia at Botany Bay. His crew asked the **Aborigines** what the strange animals were. "Kangaroo" was the reply in the aboriginal language. That means "What are you saying?" This is how we named these creatures that are only found in Australia and parts of New Guinea. There are sixty different types of kangaroos. The smallest is the muskrat kangaroo that weighs about 24 ounces (680kg), like a large grapefruit. The biggest is the red kangaroo. A "boomer," the male, can weigh 187 pounds (85kg), like the average adult man.

Caring for the Young

Kangaroos are the best-known members of a large group of mammals called marsupials. Female marsupials give birth to tiny young that are not well developed. The young crawl along the mother's belly into a pouch where they feed on milk and continue to grow. There are many kinds of marsupials, including the teddy-bear-like koala, in Australia.

Alone or in Groups

The smallest kinds of kangaroos usually live alone. Males defend a home range. The larger kangaroos, like wallabies and the red kangaroo, are social animals. They travel in groups or mobs, usually led by a male. With their powerful hind legs, kangaroos can kick savagely.

S C A L E B O X

Red kangaroo

Red kangaroo and young

Animal Athletes

The big kangaroos are good runners. They can bound along at 53 miles (88km) an hour for short distances. At full speed, a kangaroo hop may cover 28 feet (9m). But at that speed they tire quickly. At a slower pace, they can cover up to 6 feet (2m) with one hop. These animal athletes are fine swimmers and high jumpers.

The Ladybug

Imagine an insect with wings that could cover a dinner plate! That is the Queen Alexandra birdwing butterfly from New Guinea. But 300 million years ago there was a dragonfly with a wingspan three times larger.

Over 1.5 million types (species) of insects live on the Earth. Nearly a third of them are beetles. Beetles can be as tiny as the almost invisible hairy-winged dwarf beetle 0.00975 inches (0.25mm) long. Beetles can also be as big as the Goliath beetle of Africa that can be 6 to 8 inches (15 to 20cm) long and weigh as much as a small orange (3.5 ounces or 100g). But everybody's favorite beetle is the ladybug, a pretty garden insect.

Colorful and Spotty

All ladybugs have colored wing-cases. That's the shell that covers their bodies. Some are red with black spots, others are black with yellow spots or black with red spots. There are 2-spot, 7-spot, 10-spot, 14-spot, 22-spot, and 24-spot ladybugs!

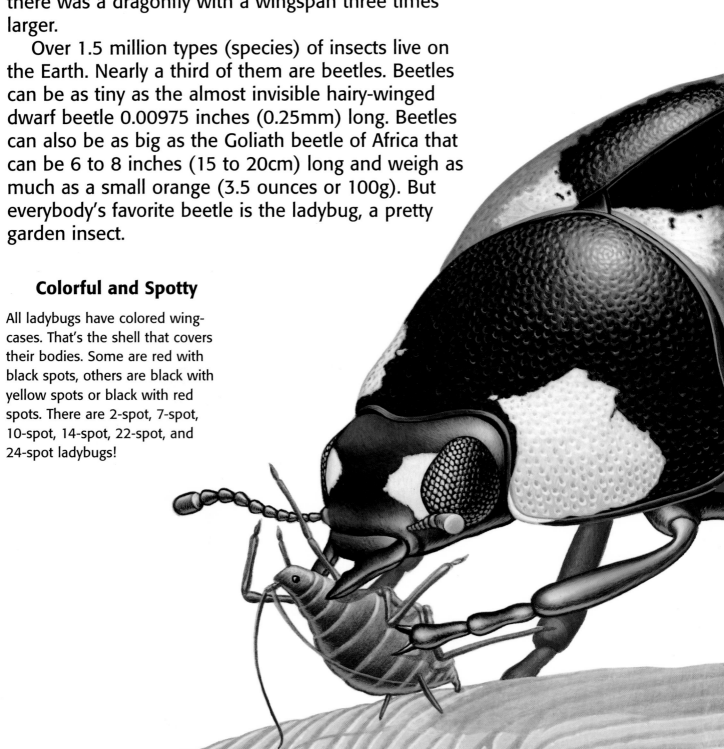

Lucky Ladybugs

Ladybugs are believed to bring good luck. They produce a yellow liquid when they are frightened. The liquid is actually blood that some ladybugs drip out through their leg joints. The blood smells or tastes bad, or is poisonous.

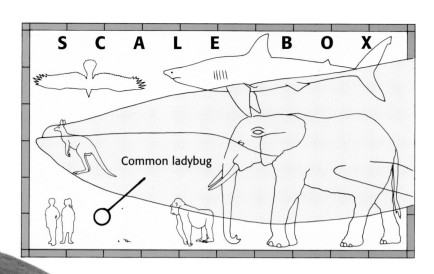

S C A L E   B O X

Common ladybug

Stay Away!

Ladybugs' bright patterns warn birds that their bodies contain poisons that are dangerous to eat.

A 7-spot ladybug sits on a human finger. The 7-spot ladybug is the most common type of ladybug.

A ladybug's wings are protected by brightly colored wing cases. The wing cases lift up when the ladybug wants to fly.

The Edible Crab

Crustaceans are animals without backbones. Instead, they have a hard outer shell that protects their bodies and allows them to move around. Other crustaceans are lobsters, shrimp, and pill bugs. Several kinds of crabs are eaten by people. In the United States, blue crabs are popular. In Europe and Britain, it's the rock crab, or edible crab, the one we can eat.

Edible crabs usually weigh about 1 to 2 pounds (0.5 to 1kg), about the same as a coconut. But they can be as heavy as 13 pounds (6kg). The body shell of adult crabs measures 5 to 6 inches (13 to 15cm) across. Like a suit of armor, the shell itself cannot grow. So from time to time the animal casts it off and grows a bigger one.

Female edible crabs are great travelers. They can walk as far as 0.6 miles (1km) a day on their many-jointed legs.

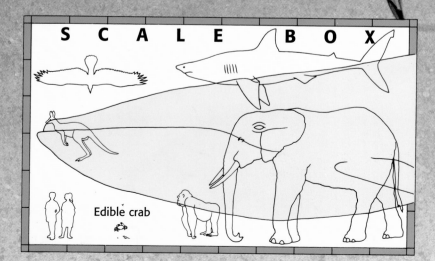

SCALE BOX

Edible crab

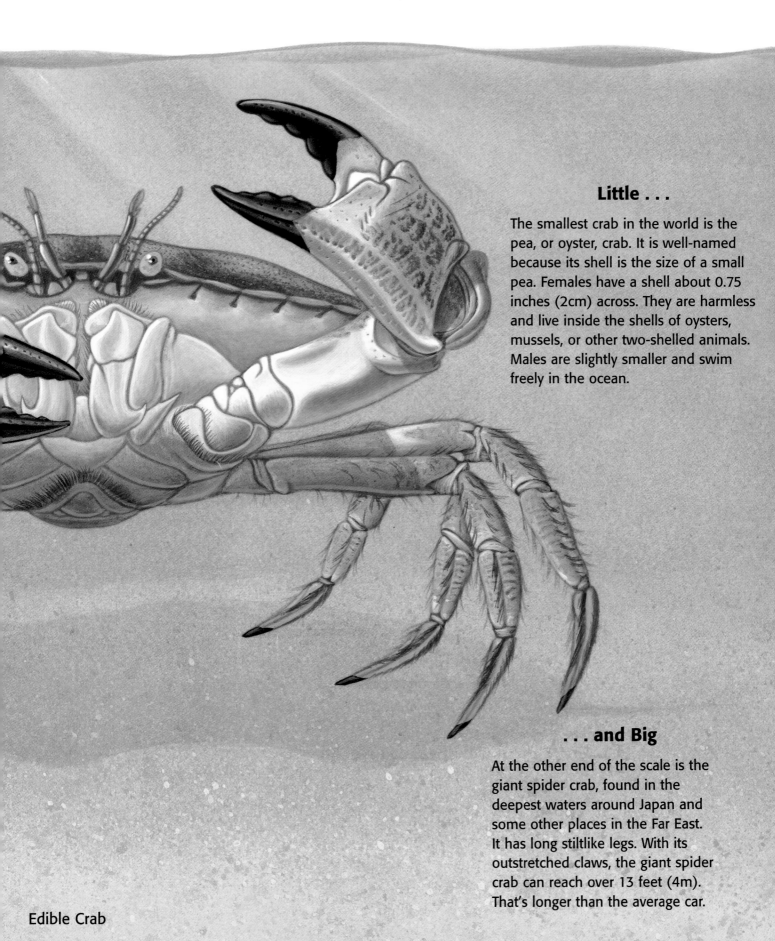

Little . . .

The smallest crab in the world is the pea, or oyster, crab. It is well-named because its shell is the size of a small pea. Females have a shell about 0.75 inches (2cm) across. They are harmless and live inside the shells of oysters, mussels, or other two-shelled animals. Males are slightly smaller and swim freely in the ocean.

. . . and Big

At the other end of the scale is the giant spider crab, found in the deepest waters around Japan and some other places in the Far East. It has long stiltlike legs. With its outstretched claws, the giant spider crab can reach over 13 feet (4m). That's longer than the average car.

Edible Crab

The King Cobra

Snakes are reptiles, which means they have dry, scaly skin and bony skeletons inside them. Their body temperature changes with their surroundings. There are about 2,400 kinds of snakes, but only 413 of them are poisonous.

The king cobra has been startled by the boy.

The well-named king cobra is the longest poisonous snake. Its head is as big as a man's hand. The king cobra feeds on other snakes. It can measure 18 feet (6m) long and weigh about 17 pounds (8kg). That is the same as an adult fox terrier. The king cobra lives in tropical Asia and the Far East.

When angry or frightened, the king cobra, like other cobras, will rear up and expand the folds of skin behind its head into a "hood" by moving some of its many ribs outward.

Snake Bites

Poisonous snake bites kill between 30,000 to 40,000 people a year. The majority of them are in India and Myanmar (Burma). Few of these deaths are caused by the bite of the king cobra. It lives deep in the jungle. Unlike some other snakes, it keeps away from towns and villages.

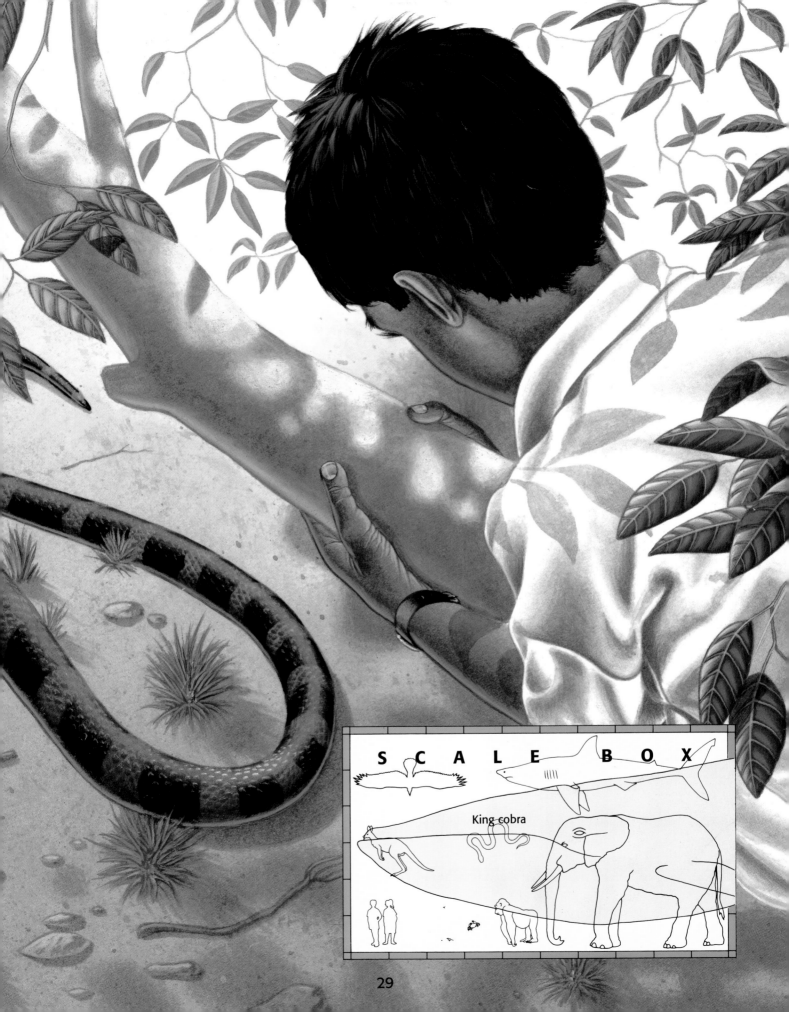

S C A L E   B O X

King cobra

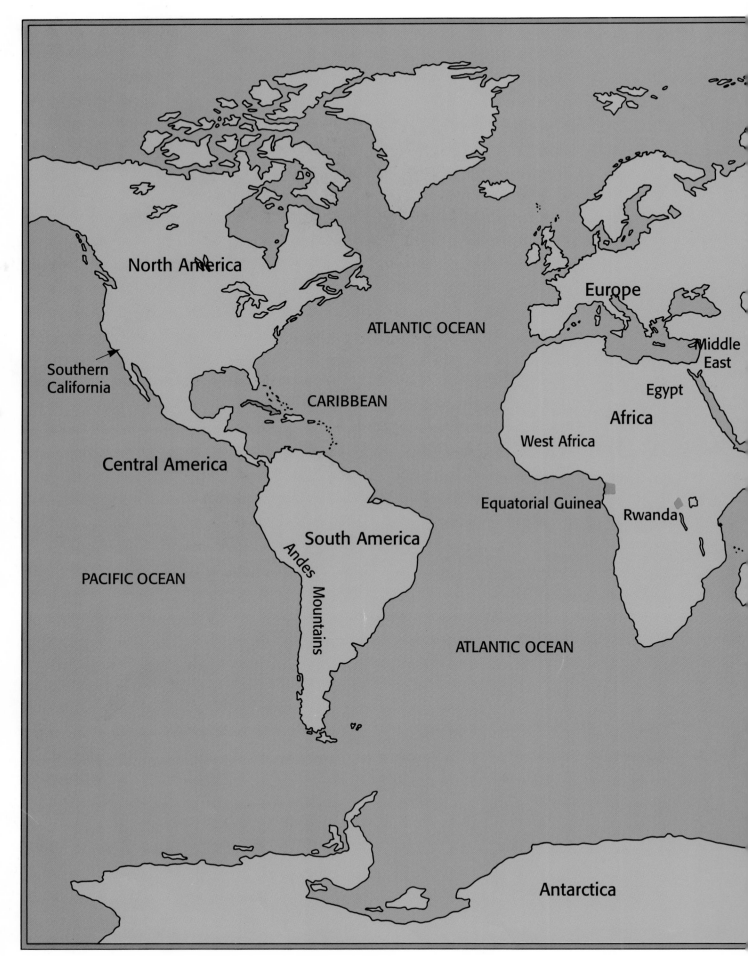

North America

Southern
California

Central America

PACIFIC OCEAN

ATLANTIC OCEAN

CARIBBEAN

South America

Andes Mountains

ATLANTIC OCEAN

Europe

Middle
East

Egypt

Africa

West Africa

Equatorial Guinea

Rwanda

Antarctica

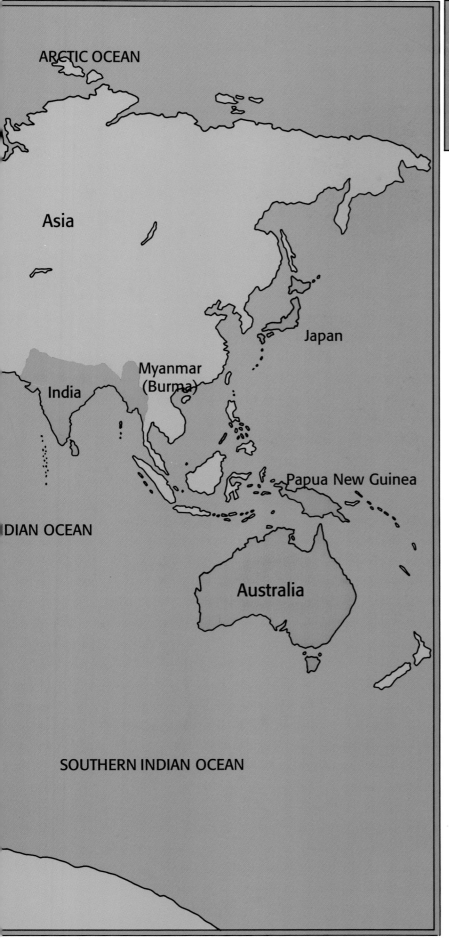

ARCTIC OCEAN

Asia

India

Myanmar
(Burma)

Japan

Papua New Guinea

IDIAN OCEAN

Australia

SOUTHERN INDIAN OCEAN

World Map

The areas highlighted in red on this map will help you to find some of the places mentioned in the book.

Glossary

Aborigines The original native inhabitants of a country. The original 500 tribal groups.

breeding grounds Areas of the world where certain animals go to have babies.

kidneys Kidneys help your body to function healthily by producing urine and getting rid of water. You have a left and a right kidney near your spine at the small of your back.

liver The liver is an organ that can be found just under your ribs. It acts as a chemical factory for the body helping to process food and destroy waste products.

mammal Warm-blooded, hairy animals that feed their young on their own milk. Human beings are mammals.

migrate Some animals move to different places in the world to breed and feed during certain seasons.

nerves Nerves are tissues composed of cells that carry electrical messages around the body.

organs Organs are collections of tissues (which are collections of cells). Organs do one or more specific jobs in the body.

sinuses Air spaces inside the bones of the skull.

Sumo A popular Japanese wrestling sport played by very large and heavy men.

tissue A collection of cells of the same kind (like muscle or bone). Different kinds of tissue make up organs and other parts of the body.

Index

A **bold** number shows the entry is illustrated on that page. A word in **bold** is in the glossary on page 31.

A
Aborigines 22, 31
Africa 9, 21, 24, 31
albatross 13
amphibians 15
Andes Mountains 12
Atlantic Ocean 30-31
Australia 22, 31

B
backbone 3
beetle **24**, **25**
birds of prey **12-13**
breathing 15
breeding grounds 7, 31
butterfly 24

C
Caribbean 11, 14, **30**
cells 3
Central America 14, **30**
cobra **28-29**
communication 16, 21
condor **12-13**
crab **26-27**
crustaceans 26

D
dinosaurs 16
dragonfly 24

E
egg 15, 17
elephant **16-17**
endangered species 7, 9
eye 3

F
feeding 6, 9, 11, 21
frog **14-15**
fish 18

G
gills 15
gorilla **8-9**

H
heart 3
hummingbird **10-11**

I
India 28, **31**
Indian Ocean 7, **31**
insects 24

J
Japan 27, **31**

K
kangaroo **22-23**
kidneys 3, 31
koala 22

L
ladybug **24-25**
language 21
liver 3, 31
lungs 15

M
mammals 6, 21, 22, 31
marsupials **22-23**
measuring length and weight 4-5
migration 7, 31
muscles 3

N
nerves 3, 31

New Guinea 22, 24, **31**
North America 12, **30**

O
organs 3, 31
oxygen 15

P
poison 14-15, 25, 28

R
reptile 16, **28**

S
scale 5
shark 7, **18-19**
shrew **20-21**
sinuses 17, 31
size
    advantages and
    disadvantages of 4
snake **28-29**
South America 11, 12, 14, **30**
Southern Indian Ocean 6, **30-31**
strength 3, 8, 17, 22
Sumo 8, 31

T
tadpole 15
tissue 3, 31
toad 15

V
vulture **12-13**

W
wallaby 22
whales **6-7**
wings 13, **24-25**